Instruction Manual for the Platinum Printing Process

1st Edition

Thomas John Shillea

Design:
Lynell Wilcha
Lynell Wilcha Graphic Design Studio

Design Production:
Marsha Bailey
Lynell Wilcha Graphic Design Studio

Photography:
Thomas John Shillea

Typography:
PHP Typography

Printing:
Indian Valley Printing Limited

ISBN: 0-9616925-0-2

About the cover logo:

In 1777, a Swedish scientist discovered that Platinum was a distinct metal, separate from silver and gold. In order to distinguish the uniqueness of the metal, he gave it the name Platinum and created a symbol for the new metal. His new symbol was based upon a combination of the existing symbols for silver ☽, and for gold ⊙.

*This book is dedicated to Sylvester,
Genevieve, Lucy and Lynell.*

Table of Contents

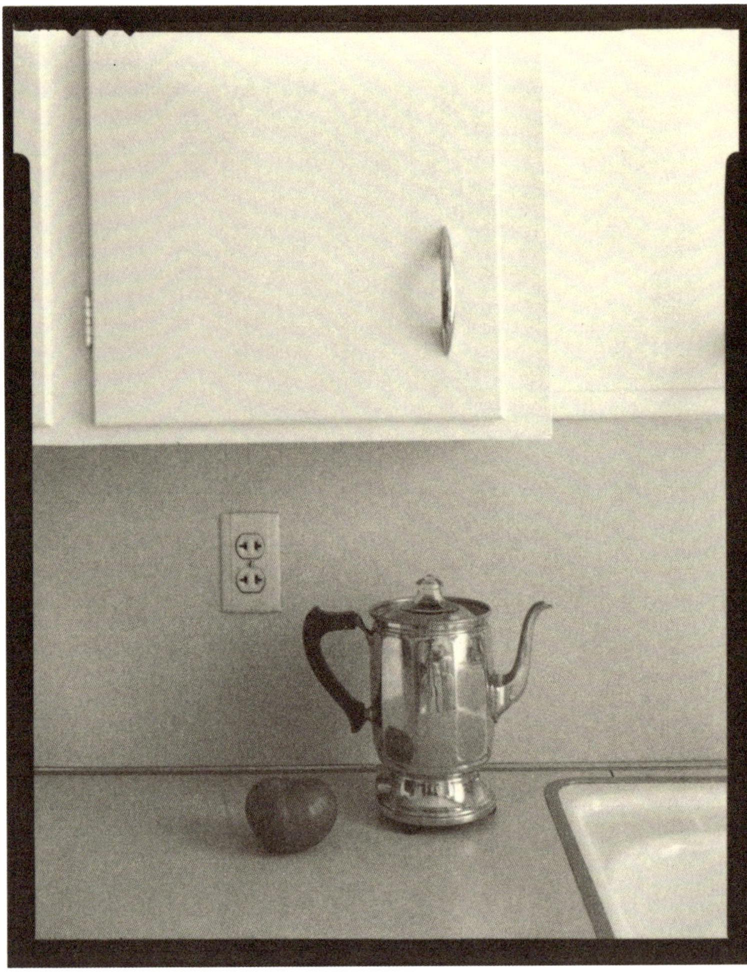

A duotone reproduction from an original Platinum Print.

1 Acknowledgements

For their assistance and guidance in preparing the information in this book, I would like to thank: The scientists and staff at Johnson Matthey in West Deptford, New Jersey, especially Dr. William Tamblyn, Technical Manager, Chemicals and Catalysts; Mr. George Cominsky, Manager; Dr. Rebecca Hanckel, Senior Chemist and Ms. Jean Michelin. Without their constant support, advice and good counsel, this work could not have been completed. Mr. Joseph Povey, also of Johnson Matthey, whose interest in the aesthetics and historical importance of the Platinum Print led to the establishment of the Johnson Matthey Platinum Print Collection; Mr. Charles Isaacs, for acting as a consultant to Mr. Povey on the history and aesthetics of the Platinum Print, and for being a valuable source of sound advice to the author; photographers Paul Runyon, Michael Furman and Victoria Satterthwaite, for their encouragement and support; Mr. James Davis and Ms. Sue Erb of Blue Mountain Photo Supply, for their generous contribution of many of the processing chemicals used to conduct the experiments and tests recorded; Associate Professor Owen Butler, Professor Richard Zakia and Mr. Thomas Iten, Director, from the School of Photographic Arts and Sciences at The Rochester Institute of Technology, for their suggestions and criticism and their continued dedication to photographic aesthetics and education; Ms. Lynell Wilcha, for her sensitivity and commitment in designing and supervising production; Ms. Holly Maxson, for sharing her expertise and information on paper conservation which is included in Appendix C; and finally to photographer John Hafey, who in 1978 introduced me to the beauty of the Platinotype Process and who helped me write ***The Platinum Print***, published in 1979.

Thomas John Shillea
Philadelphia, Pennsylvania
October 1986

2 Introduction

From the year of its commercial introduction in 1873, through the second decade of the 20th century, the Platinotype Process was used by the premier photographers of the era. It was highly favored by these photographers not only for its archival permanence, but also because of the dynamic tonal range (from rich, velvety blacks through a seemingly endless range of mid-tones, to soft and delicate whites) produced in the Platinum Print.

Many of the great photographs made by Peter Henry Emerson, Frederick Evans, Alfred Stieglitz, Gertrude Kasebier, Edward Steichen, and Paul Strand (to name a few) were printed on platinum paper.

For various reasons, the Platinotype Process waned in general popularity after 1930*. However, a group of dedicated photographers, represented by such individuals as Karl Struss, Laura Gilpin, Immogen Cunningham and Paul Outerbridge continued making Platinum Prints on hand-coated paper right up until the 1960's. The work of these and other photographers allowed a new generation to experience the beauty and power of the Platinum Print, thereby encouraging a renaissance in the Platinotype Process.

From the early 1970's to the present, more and more discerning photographers are rediscovering the pleasure of working in this process which Alfred Stieglitz once called "the prince of photographic printing processes".

From the time that the Platinotype Process was invented by William Willis in 1873, Johnson Matthey has been supplying photographers and photographic manufacturers with the platinum metal compounds necessary to make platinum prints. As part of this involvement, Johnson Matthey recognized the need to provide contemporary photographers with a detailed, practical and modernized manual for the safe and successful practice of the Platinotype Process. In addition, an effort was made to update the scientific terminology, as well as the theory of the process, incorporating the results of recent scientific research in the areas of photochemistry and archival processing for the permanence of the Platinum Print. Johnson Matthey has proudly supported and encouraged the author in the production of this fine manual.

*See *"The Platinum Print"*.

3 Platinotype Instruction Manual

This instruction manual will introduce the novice to the simplicity of the technique and the beauty of the Platinotype Process, and make available to experienced photographers a contemporary source of information to add to their present knowledge.

The manual is not intended to be the definitive statement on the techniques of the Platinotype Process. As with any creative process that offers personal manipulation as one of its strong points, the Platinotype Process is as varied as the individuals working with the process. It is the intention of the author that this Instruction Manual will help to perpetuate this atmosphere of experimentation and creativity.

The Platinotype Process has been in use for more than 100 years. During this time, many instruction manuals have been written as guides in making Platinum Prints. A listing of the classic and contemporary writings on the subject is presented in the bibliography of this publication. This will be of particular importance to those individuals interested in further readings and as an aid for research.

Of all the vintage photographic processes, the Platinotype Process is one of the simplest in concept and technique. The photographer need only a general understanding of the chemicals used in the process and a minimum amount of hardware and materials in order to begin his work. It is even possible to make Platinum Prints without the use of a darkroom.

4 Safe Handling of Chemicals

Intelligent and proper handling of all photographic chemicals will guarantee safety, consistent results and, of course, will save the worker time and money.

It is wise to consider any chemical as potentially hazardous and to handle it with caution. This point cannot be emphasized strongly enough. These precautions include wearing protective gloves and safety glasses when mixing and using the chemicals. If contact with the skin should occur, rinse the exposed areas with large amounts of water. It is also advisable to wear a darkroom apron to protect your clothes from staining, to handle the prints in solution with wooden or plastic tongs, and always work in a well ventilated space. Never store any photographic chemicals near food or in areas accessible to children. Make sure that all storage containers are properly labeled, and avoid contaminating one chemical with another.

Always store chemicals in a cool, dry place in tightly capped containers. As some of the chemicals used in the Platinotype Process are quite light sensitive, they must be stored in light resistant containers.

It is not advised to store any of the Platinotype chemicals in a refrigerator or at low temperatures. This may cause some of the chemicals to precipitate. If this should occur, gently warm the solution to about 90 degrees F (32 Degrees C) and mix until the solid redissolves.

For more detailed safety data see Appendix F.

Equipment Necessary for the Platinotype Process

Darkroom Equipment

The following is a comprehensive list of basic materials and equipment required to make a Platinum or Palladium Print:

1. A contact printing frame, at least one size larger than the negative to be printed in the frame. The classic hinged-back frame is best. This will allow you to inspect the image as it prints out during exposure. The Premier Company makes a suitable line of frames that are available at most photo dealers.

2. A small hairdryer, with settings for low and high heat.

3. A soft brush, without any metal parts. The classic type is the Japanese Hake

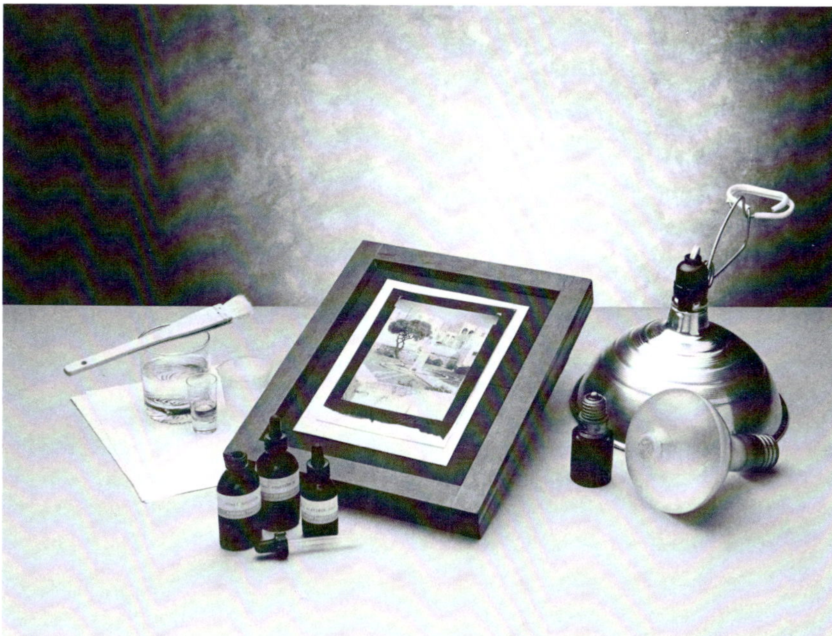

Basic chemicals and equipment necessary for the platinum printing process.

(pronounced as in "bake") brush. However, a common foam brush, available at paint stores, will suffice for the initial printing experiments. The brush should be at least 1½" in width and should not be used for any other purpose than coating the sensitizer on the paper.

4. Two clean glass containers, one to wash out your brush after coating the paper, and one in which to mix the sensitizer solution.

5. A clean piece of ordinary window glass or Plexiglass on which to tape your paper during coating. It should be at least two times larger than your paper size.

6. Five clean processing trays. The trays should be either glass or plastic (not metal) and should be reserved only for platinum processing chemicals.

7. A Kodak Tray Syphon washer or similar type device for washing your prints after processing.

8. An incandescent light source for your darkroom or coating area. Ideally, no more than a 60 watt bulb placed at least six feet from your work area. Do not use a fluorescent light in your work area as it emits some UV radiation and may partially fog (expose) your paper during coating and drying.

9. Two or three gallons of distilled water.

10. A clean plastic gallon container to store your developer solution.

11. A pair of scissors, a ruler, lead pencils, masking tape.

12. Disposable plastic gloves, an apron, either plastic or cloth.

13. Plastic or wooden tongs for print processing.

Light Sources for Exposing the Sensitized Paper

The least expensive light source for the Platinotype Process is sunlight. It is rich in UV radiation and, of course, was the type of light used by the pioneers of the Platinotype Process. However, the main disadvantage of printing with sunlight is that its UV output changes hourly and with the seasons. This fluctuation makes reproducible printing exposure times almost impossible. It is therefore advisable to use some artificial light source which produces UV radiation for exposing your print.

The most common choices are:
1. 275 Watt Sunlamp
2. Instant Mercury Vapor
3. Carbon Arc
4. Fluorescent Black-Light Bulbs

It is important to remember when selecting a light source for exposing the print that it is not the intensity of the light source that is important, but rather the amount of UV radiation the light source will produce.

Standard 275 watt sunlamp bulbs are manufactured with a mogul base for safety purposes. This mogul base is larger than the normal light bulb base and will not fit into an average size bulb receptacle. You will therefore need a mogul base adaptor to fit the sunlamp into the clamp light housing. These mogul base adaptors are available at lighting and hardware stores.

A 275 watt sunlamp will offer hundreds of hours of consistent exposure for your Platinum Printing. With the proper density and contrast negative, your printing (exposure) time with two 275 watt sunlamp bulb will be approximately 15 to 20 minutes.

If you elect to print with a sunlamp bulb, there are some things to consider:

1. Shield your eyes from looking directly into the light source for prolonged periods of time, just as you would from the sun.

2. Turn the light on at least 10 minutes before you intend to make your exposure to allow the bulb to reach its peak UV output.

3. Leave the light turned on during the entire printing session. If the lamp is turned off it will require a 15 minute cool down time before it can be turned on again.

Theory of the Platinotype Process

The Platinum and Palladium Salts used in the Platinotype Process are only slightly light sensitive. Therefore, the process requires the intervention of some other, more photosensitive chemical reagent. The most consistent results are obtained using Ferric Oxalate [iron (III) oxalate]. This compound, when exposed to UV radiation, produces a visible image as the Ferric Oxalate is reduced to the ferrous state. This initial visible image is referred to as the provisional image.The Ferrous Oxalate then chemically reduces the platinum or palladium salt to the pure metallic state when the exposed print is placed in a suitable developer solution.

This ability of the ferrous compound to reduce Platinum, or any Platinum group metal, is the foundation of the Platinotype Process. As stated in Appendix D, pages 56 and 57 of this manual, there are three chemical solutions used to make the light sensitive formula. Both the A and B solutions contain Ferric Oxalate. However, the B solution also contains Potassium Chlorate which allows for contrast control in the Platinotype Process.

When Potassium Chlorate is added to the sensitizer, it acts as an oxidizing agent. The effect of the Potassium Chlorate will be seen in the heightened brilliancy of the finished print. Another effect of the Potassium Chlorate is to reduce the sensitivity of the paper, thereby requiring longer exposure times. An excess of Potassium Chlorate in the sensitizer may also cause granularity in the finished print. In simpler terms, the Potassium Chlorate is added when more contrast is needed in the print, as it tends to undo the effect of UV radiation on the Ferric Oxalate, thereby cleaning up the highlights in the print. In this way, it is like chemical dodging.

8

Processing Chemicals

One of the unique aspects of the Platinotype Process is that it operates as both a printing-out and developing-out process. Once the provisional image of Ferrous Oxalate has been formed by exposure to UV radiation, the print is then placed in a suitable developer solution and the Ferrous Oxalate is replaced by the platinum metal. There are approximately 16 different chemical solutions that can be used to develop the provisional image, one of which is distilled water. The classic solution used by William Willis when he invented the process in 1873 was Potassium Oxalate. Two other solutions which function in similar fashion are Sodium Acetate and Sodium Citrate. Once the formation of the platinum image is completed by development, any remaining Ferric Oxalate, which is still light sensitive, must be removed from the paper to prevent further darkening of the image. This is simply achieved by etching the print in three successive baths of a weak acid solution. The acids most commonly used are Hydrochloric and Citric. Of these, Citric Acid is the safest to use, as it is non-toxic and non-corrosive to the skin.

Sensitizer Formulas

Following is a list of formulas using the A, B and C solutions to make a range of contrast grade sensitizers to fit the density range of the negative to be printed. The amounts represented are in drops and are sufficient for single coating a 4 × 5 inch area on a piece of paper. To coat an 8 × 10 inch area, simply double the amount. Naturally, it is important that all three droppers are the same size.

Formula 1

Solution A = 0 drops
Solution B = 8 drops
Solution C = 10 drops

This formula is designed to produce "normal" prints from very soft negatives (density range of approximately 0.60 to 0.70). This type of negative would require a grade 5 or 6 silver halide paper.

Formula 2

Solution A = 2 drops
Solution B = 6 drops
Solution C = 10 drops

Formula 2 is designed to produce "normal" prints from soft negatives (density range of approximately 0.80 to 0.90). This type of negative would require a grade 3 or 4 silver halide paper.

Formula 3

Solution A = 5 drops
Solution B = 3 drops
Solution C = 10 drops

Formula 3 is designed to produce "normal" prints from average negatives (density range of approximately 1.00 to 1.10). This type of negative would require a grade 2 or 3 silver halide paper.

Formula 4

Solution A = 6 drops
Solution B = 2 drops
Solution C = 10 drops

Formula 4 is designed to produce "normal" prints from contrasty negatives (density range of approximately 1.20 to 1.30). This type of negative would require a grade 1 silver halide paper.

Formula 5

Solution A = 8 drops
Solution B = 0 drops
Solution C = 10 drops

Formula 5 is designed to produce "normal" prints from extremely dense and contrasty negatives (density range above 1.40 to 1.60). This type of negative would require a grade 0 silver halide paper.

The total number of drops of the A, B and C solutions may have to be increased depending upon the absorbancy of the paper being coated.

It is also possible to make intermediate contrast grades by making even more subtle adjustments in the amount of A and B solutions used in a sensitizer formula. The addition or subtraction of just one drop of solution B can make a significant difference in the appearance of the finished print. Even more tonal and contrast control is possible by double coating the paper. The result is usually a richer and more contrasty print than obtained from single coating the paper. Bear in mind that a greater number of drops of B solution will decrease the sensitivity of the paper and increase the necessary exposure time. In addition, this may give a harsh appearance to the print with an excess of granularity in the image.

10 Making A Platinum Print

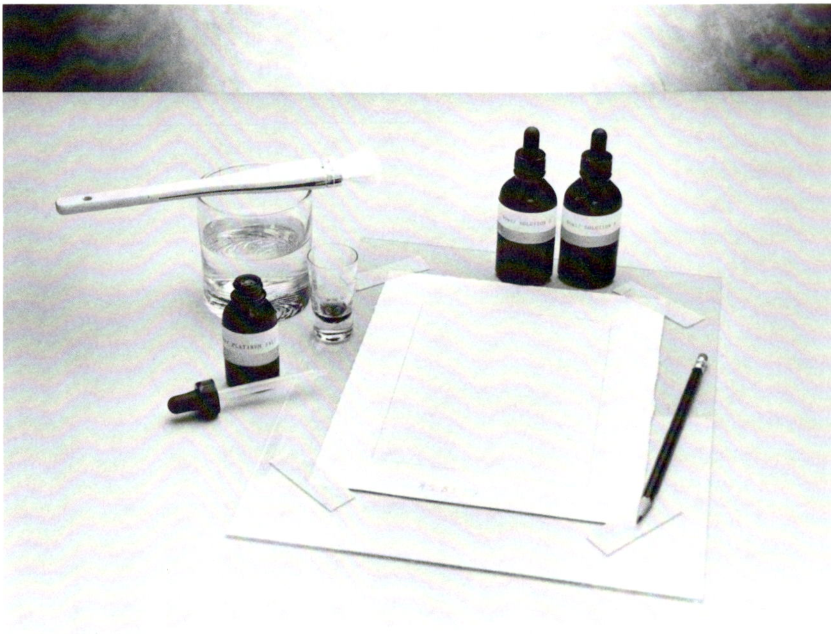

Figure 1. Chemicals and materials necessary to sensitize a sheet of paper.

At this stage of the procedure, the A, B and C solutions should be mixed and ready to use. See Appendix D in this manual for the chemical formulas for mixing the A, B and C solutions. For all intents and purposes, the Platinum and Palladium salts will work identically in the sensitizer formula. It is also possible to mix proportions of the Platinum and Palladium together to make the C solution. This procedure offers certain attractive advantages:

1. Palladium will warm-up the image color of the platinum print.
2. As Palladium is less expensive than Platinum it offers a sensible method of reducing the cost to print.
3. If the ratio of Platinum to Palladium is at least 1:4, the image permanence of the print will be equivalent to a 100% platinum metal.

Developer Solution and Acid Clearing Baths

The developer solution should be mixed in advance of the printing session. The mixing formula for the standard developer compounds is 454 grams (1 pound) to 50 ounces (1478 ml) of distilled water. The developer can be mixed with the water at room temperature.

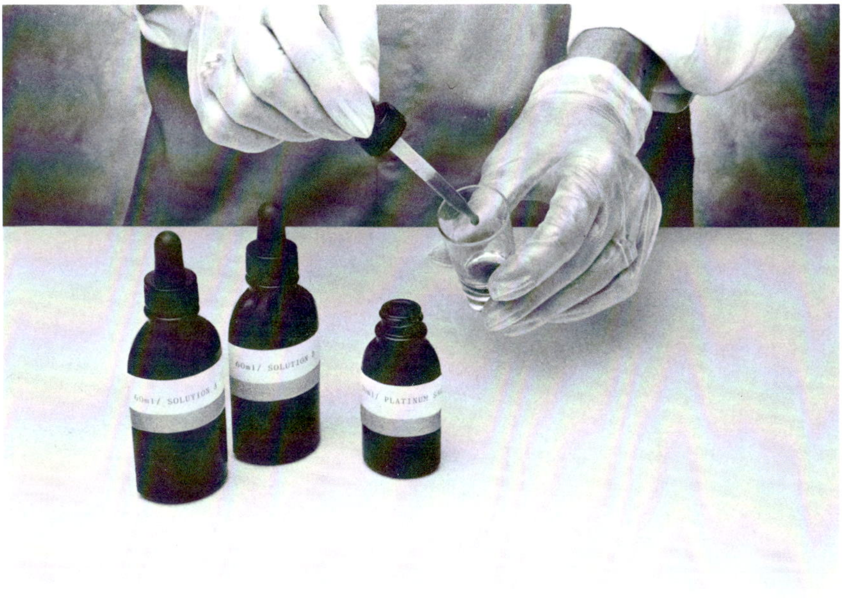

Figure 2. Measuring out the drops of A, B and C chemicals to make the sensitizer solution.

The three trays of citric acid are mixed with ordinary tap water in the proportion of 60 ounces water (1478 ml) to 30.0 grams of Citric Acid Powder. A simple method of determining the correct amount of Citric Acid to add to the 50 ounces of water is to fill a plastic graduate up to the 30 ml line with the Citric Acid Powder.

Sensitizing the Paper

Cut your paper at least 2 inches larger in both length and width than the negative size you will be printing. Before sensitizing the individual sheet, inspect the paper via transmitted light to see if there are any large specks of impurities in their fibers. Use the cleanest piece of paper you have. Do not touch the surface of the paper with your fingers or any other material. Secure your paper to a sheet of glass (or plexiglass) by all four corners with masking tape. Pencil in the area to be covered with the sensitizer solution. Simply tracing around the negative sleeve containing the negative will suffice for this purpose.

Always wear protective gloves, and an apron while coating your paper. Fill a clean

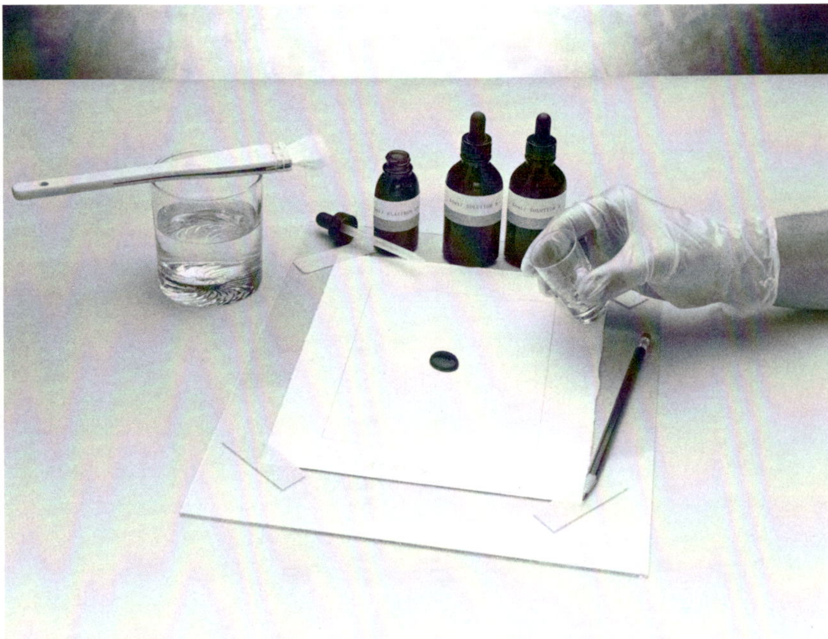

Figure 3. Pouring the sensitizer solution onto the sheet of paper.

glass with distilled water and rinse out the coating brush. Squeeze out all the excess water from the brush; it should be only damp, not wet. Select the sensitizer formula appropriate for the negative to be printed. Measure out the drops of A, B and C solutions in a small glass and gently swirl the glass to mix the solutions together.

Pour the sensitizer onto the center of your paper and immediately, but with gentle pressure, brush the sensitizer over area of your paper. Brush in only one direction. Allow the sensitizer to soak into your paper for at least two minutes, while the paper is still taped to the glass.

Drying the Paper

Hang the paper up in your darkroom and allow it to airdry for another 8-10 minutes. While the paper is drying, rinse out your brush and sensitizer glass with distilled water. Repeat this procedure after each paper coating. After the paper has airdried for 10 minutes, blow-dry the front and back of the paper with your hair dryer set on either warm or style, not on hot. Keep the heat source at least 6 inches

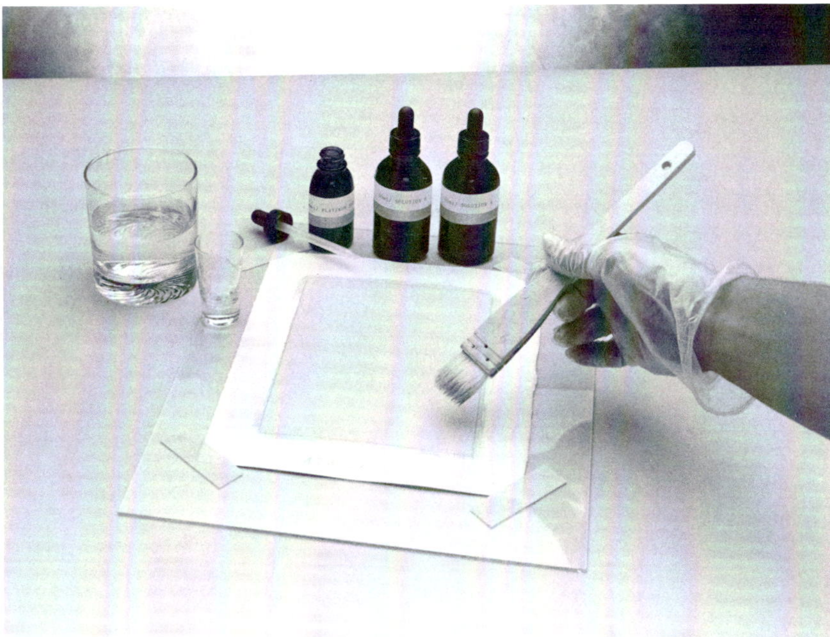

Figure 4. Coating the paper with the sensitizer solution using a hake brush.

away from the paper during drying. Blow dry each side of the paper for one to two minutes.

The drying procedure is important to the success of the finished print because the sensitizer must have sufficient time to soak into the paper, so that it will be held firmly in place during chemical processing. If insufficient drying occurs, the sensitizer will remain on the surface of the paper and two results will occur:

1. The finished print will lack the desired contrast, due to the paper being damp during the exposure.
2. Some of the Platinum metal will float off the paper during processing, which will also weaken the intensity and tonal range of the finished print.

If too long a drying time is allowed, the sensitizer will sink too far into the paper fiber and the finished print will appear flat because the paper texture will scatter too much of the light reflecting from the paper surface. In addition, too much heat may decompose the sensitizer which will, in turn, cause uneven tonal densities in the finished print.

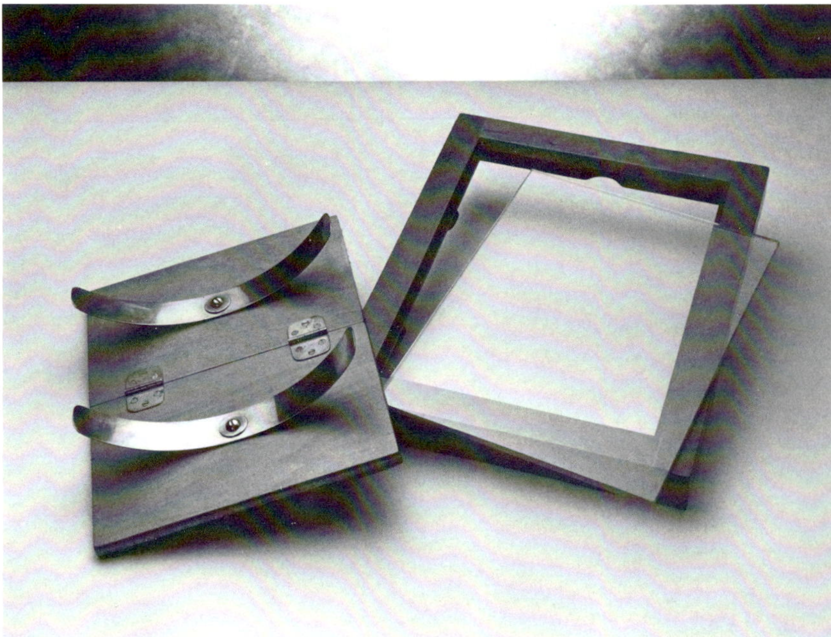

Figure 5. The classic, hinged-back contact printing frame.

It is more likely that you will err in favor of the paper being too damp as opposed to too dry. In other words, it is better to overdry the paper than to print the paper while it is still damp. Unless you are very careless, it is quite difficult to scorch the surface of the paper during blow-drying. When dried, the paper should crackle when slightly bent by hand. Another method of determining the proper drying of the paper is to place the back of the dried paper against the back side of your hand. The paper should feel only slightly damp, not cold or wet. For superior results, it is recommended to expose the paper immediately after drying. Once the paper is dry, it is quite sensitive to UV radiation and must be shielded from such light sources. The paper can be safely handled under the incandescent light in your work area.

How the Contact Printing Frame Works

The Platinotype Process is a contact printing process which requires the use of a hinged-back contact printing frame to keep the negative in tight contact with the sensitized paper during exposure. This tight contact is necessary to guarantee sharpness and details in the finished print.

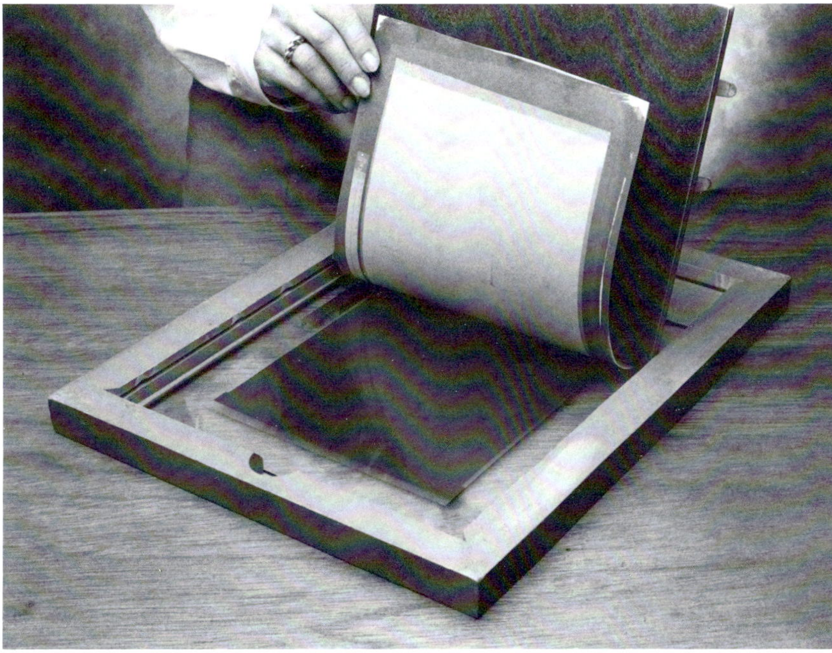

Figure 6. Inspecting the printed-out image in the contact printing frame.

Another function of the contact printing frame is to allow for visual inspection of the provisional image as it prints-out during exposure to light. It is this visual inspection that enables the photographer to determine the correct exposure time for the print.

The classic hinged-back printing frame is designed with a back section which is divided into two pieces held together by hinges. During exposure, one half of the hinged-back can be released by freeing the pressure clips holding it in place. This allows for half of the back of the frame to be opened, revealing the exposed print for visual inspection. Because one of the pressure clips remains in place, the contact and registration of the negative and paper is not lost or disturbed during visual inspection.

Assembling the Sensitized Paper and the Negative in the Printing Frame

After sensitizing and drying the paper, disassemble the printing frame and clean both sides of the glass. Use Dust-Off or an anti-static cloth to clean the negative. Place the sensitized paper, face-up, on the felt side of the hinged-back of the frame.

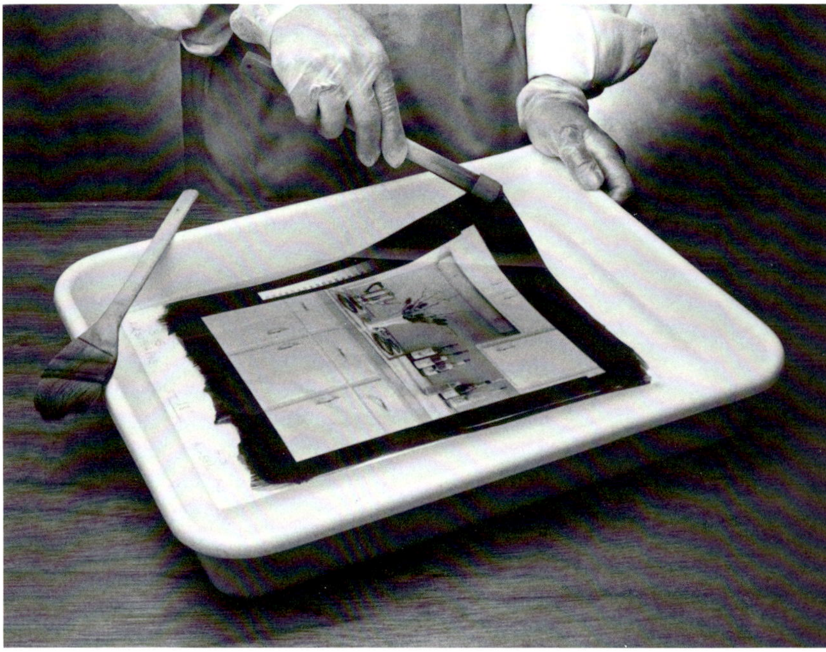

Figure 7. Developing the exposed paper in a solution of Potassium Oxalate.

On top of the sensitized paper, place a clean sheet of clear acetate, ideally no more than .003 to .005 mm thick. (This material can be purchased at an art supply store.) Place the negative, emulsion side down on top of the clear acetate, and finally place the sheet of glass on top of the negative. This whole section is then returned to the frame and held in place by the metal pressure clips.

The use of clear acetate is not a necessity, but will serve to protect any valuable negatives from possible chemical staining due to being printed in contact with a damp piece of sensitized paper.

Place the assembled frame, containing the paper and negative, under the exposing light source. If a sunlamp is used, keep the bulb(s) at least 20 inches above the frame during exposure. Make a notation of the time at the beginning of the exposure, and note the duration of the exposure time for future reference when printing a particular negative. After exposing the print, remove the paper and negative from the contact printing frame and process the print as described in the following section.

17

Processing the Exposed Print

Development

Immediately after exposing your print, place it face-up into a tray containing the developer solution. Always use plastic or wooden tongs to handle the print during processing. The developer solution may be used at room temperature or may be heated to a temperature of 90/100 (32C/37C) degrees F. This will make the solution act more vigorously. Development will occur almost immediately upon placing the exposed print into the heated developer. This increased activity of the heated developer will allow for shorter exposure time for the provisional image.

One word of caution when working with a heated developer solution: because development occurs almost immediately, it is important to submerge the print quickly and evenly into the developer solution to avoid streaking and uneven development. Also watch for air bubbles that may form on the surface of the print. If this should occur, brush over the areas immediately with a soft-hair brush. Whether using the developer at room temperature or heated, allow the print to remain in the developer solution for at least two minutes. You cannot overdevelop a Platinum Print.

Etching

After development, remove the print from the tray and allow the excess developer to drain back into the tray. Then transfer the print to the first acid-bath tray. Allow the print to etch in the acid-bath for five minutes with intermittent agitation. Remove the print from the first acid-bath tray and allow it to drain. Place it in the second acid-bath tray and repeat the procedure.

After five minutes, remove the print from the second acid-bath and place it in the final acid-bath for five minutes. The acid solution in the first tray will soon become cloudy and yellowed in appearance. This is due to any remaining Ferric Oxalate being etched from the paper. When this occurs, discard the solution in the first acid-bath tray, move the second and third trays into the first and second positions and fill the third tray with a fresh acid-bath. The third tray should always contain a clear acid solution.

A word of caution about etching the developed print. This is a critically important part of the process and is usually passed over with casual comments in other manuals. However, if the print is not properly etched, it will darken with age and in this way can be destroyed. It is also important not to forget a print in any of the acid-baths. Air bubbles may form on the surface if the print is left to float unattended in the acid-bath. This will cause uneven etchings of the Ferric Oxalate in the paper. Extended soaking in an acid bath saturated with iron salts will most definitely stain the clean borders of the paper and may even stain the image area. It is important to keep the ratio of acid to water in the acid batch as indicated. If too strong an acid bath is used, the solutions will become super saturated with Ferric Oxalate which will cause yellowish stains on the print paper. All that is necessary to avoid these potential problems is to monitor the print while it is in the acid baths.

Washing the Finished Platinum Print

After development and etching, the Platinum Print must be washed thoroughly to remove as many acidic residues and heavy metal impurities as possible. The sensitizing solution of Ferric Oxalate and the acidic etching baths used in the printing process leave contaminants which contribute to the deterioration of the cellulose structure of the paper.

The recommended washing rate is a complete exchange of water every 5 minutes for an hour. This can be accomplished by either discarding and refilling the water bath every five minutes twelve consecutive times, or by allowing the water to flow through the bath, so that the water bath is fully refreshed every five minutes. The rate of flow can be tested before you add prints to the wash bath by putting a little water soluble food coloring in the wash bath and timing its disappearance. In addition, water at warmer temperatures increases the swelling of paper fibers and gelatin size (if used), which allows for more efficient release of contaminants. The wash water should be no colder than 68 degrees F (20 degrees C).

Water Used for Print Washing

If you plan to use tap water for the washing cycle, it is wise to have it tested for contaminants and heavy metal residues before assuming that it is harmless for your prints. It is equally harmful to wash with water which is too alkaline (pH 7.0 and above). There are companies which are equipped to test water.

Distilled water is water from which all metallic and organic impurities have been removed. Using distilled water can be advantageous for two reasons. First, it does not add contaminants to the paper and second, it tends to leach out metallic impurities (such as iron and copper) in the cellulose structure of the paper. Distilled water is readily available from local sources. If distilled water alone is used for washing, the tray refill method should be used. It is also possible to do the initial print washing with tap water and the final still bath with distilled water.

De-ionized water is also a purified form of water, but because it is passed through an ion exchange column to remove impurities, it often has an acidic pH. De-ionized water may also contain salts and minerals. Calcium, in the form of Calcium Hydroxide, is often added to the deionized water to achieve a neutral pH.

Print Finishing

After thorough washing, the print is hung up and allowed to air dry. With most papers used in the Platinotype Process, the print will return to an almost flat condition after drying. However, if the paper remains slightly curled after air drying, it can be gently pressed flat in a drymount press set at 150 degrees F. Any minor blemishes on the print surface can be spotted using standard Kodak or Spot Tone liquid dyes.

Storage of the Finished Print*

Environmental and storage conditions are as important in preserving the Platinum Print as are the processing steps used in producing the prints. High temperatures and relative humidity work together to initiate the breakdown of the cellulose molecule, which results in darkening and embrittlement of the paper. In fact, for every 10 degree C rise in temperature, the rate of deterioration doubles. Recommended storage conditions are 40-50% relative humidity and 60-70 degrees F (16-21 degrees C) and drastic fluctuations in the environment are to be avoided.

Exposure to bright intensities of light and to ultraviolet radiation also cause discoloration and degradation of the cellulose structure. No photograph or other object made from paper should ever be on permanent display, and three months is considered the maximum allowable period. Ultraviolet filtering plexiglass is available for prints which will be on display for such an extended period.

Storage within an acidic folder or within a wooden cabinet can also initiate further degradation by an acidic transfer. Prints can be safely stored in lignin-free envelopes and folders within acid-free boxes. These materials are readily available from archival sources, such as Light Impressions in Rochester, NY.

*This information supplied by Ms. Holly Maxson.

Judging Print Exposure Time

One of the advantages of the Platinotype Process is that it partially prints out the positive image. This printed out image can be visually inspected and used as a barometer to judge the exposure. This is the most common method of judging print exposure, and after some experience, can be mastered by anyone making Platinum Prints. There are several methods for judging the correct exposure time:

If you are going to visually judge the exposure, follow these suggestions:

1. Always inspect the printed out image in an area which is shielded from UV radiation.

2. Be careful when opening and closing the hinged back of the frame not to lose your negative registration. Of course, you will only be opening half of the hinged back of the printing frame and peeling back the paper to judge your exposure.

3. Continue printing until the shadow details are quite visible and the mid-tone details are slightly visible in the provisional image. Do not continue printing until the highlight details are visible. If you do, the print will be badly overexposed.

A second method of judging exposure time is to make a test-print.
The procedure is as follows:

1. When coating your paper, cover an extra area that you can cut off to make your test print.

2. Cut this extra strip from your dried paper and place it in the printing frame under your negative. Make sure to include a highlight and shadow area from the negative on the test-print.

3. Print out the provisional image until you think it has received the proper exposure. Do this by visually judging the provisional image.

4. Process the test print, etch it in one acid bath and then wash for one minute and squeegee dry. A Platinum Print dries down about ½ to 1 full stop; therefore, do not attempt to make any judgments about your print while it is still wet.

5. If the test print looks right, then simply expose the full sheet of paper to your negative for the same time. If it looks over or underexposed, then estimate the difference in stops and make a judgment on how much to increase or decrease your printing time. A stop of exposure represents either a doubling or cutting in half of the original time. Therefore, if your test print looks underexposed by a half-a-stop, and your test-print exposure time was 10 minutes, then expose your next print for 15 minutes (a full-stop more would have required 20 minutes). Therefore, a half-stop more exposure would require 15 minutes of exposure.

Making test-prints during your initial experiments with the Platinotype Process will save you time and money and give you good experience on how the process works. Eventually, you will become quite capable of selecting the correct sensitizer formula and printing time for each of your negatives.

Any photographer familiar with the concepts of sensitometry and who has access to a densitometer will be able to work out a system of exposure based upon the relationship between the sensitizer formula and the density range of the negative being printed. This procedure calls for simultaneously printing a #2 Kodak Step-

Tablet with the negative. A clear explanation of this technique, including an exposure chart, is contained in *The Keepers of Light*, listed in the bibliography.

The author has used all three methods and finds the visual judgment of the provisional image, based upon experience, to be the most satisfying. However, it is up to the individual to design the most convenient and satisfactory system.

Regardless of which of the methods of judging print exposure is used, it is important to keep all the variables as constant as possible. Sensible and consistent work habits will guarantee a higher level of success. It is also advisable to practice the techniques of Platinotype with Palladium until familiar with the process, and then begin working with Platinum.

How to Interpret the Test Print

In order to obtain the maximum effectiveness from this exquisite photographic process, two aspects of the printing technique must be controlled. These are:

1. Selecting the correct sensitizer formula (mixture of the A, B and C solutions) for the density/contrast negative to be printed.

2. Judging the correct exposure time.

Methods of selecting the sensitizer formula and judging the print exposure time have already been discussed. Once the test print has been developed, properly etched to remove any Ferric Oxalate, washed and dried, it can be interpreted according to the following criteria:

1. Just as with a black-and-white silver print, the rule of thumb is to expose the print for the highlight detail in the negative. Therefore, if the dried Platinum Print lacks the required highlight detail and is also weak in the shadows, then the print was underexposed.

2. If the dried Platinum Print contains a good highlight detail, but the shadow areas of the print are too dark, then the sensitizer formula was too contrasty for the negative printed.

3. If the dried Platinum Print is generally gray in appearance, the highlights are clearly detailed but too dark, and the shadows are also too dark, then the sensitizer formula was not contrasty enough for the negative printed.

The solution to problem #1 is obvious enough: simply increase the exposure time for the next print. The solution for problems #2 and #3 is to either decrease or increase the number of drops in the B solution in the original sensitizer formula. Remember to compensate for the change in the amount of B with a corresponding change in the amount of solution A so that the sensitizer formula will always have the same number of drops. For example, if the initial sensitizer formula had contained:

Solution A = 5
Solution B = 3
Solution C = 10
and after inspection of the test print a decision is made to increase the contrast of the print by adding two drops of solution B, then the new sensitizer formula would be:

Solution A = 3
Solution B = 5
Solution C = 10

Remember that changing the amount of B in the sensitizer formula will also require a compensation of the print exposure time. As a starting reference, use the following exposure adjusting formula:

For every drop of B solution added to or subtracted from the original sensitizer formula, either increase or decrease, respectively, the exposure time of the print by 10-15%.

Finally, as in any creative process, there is no right or wrong way to print a negative, only the best way suited to express the strongest interpretation of the subject.

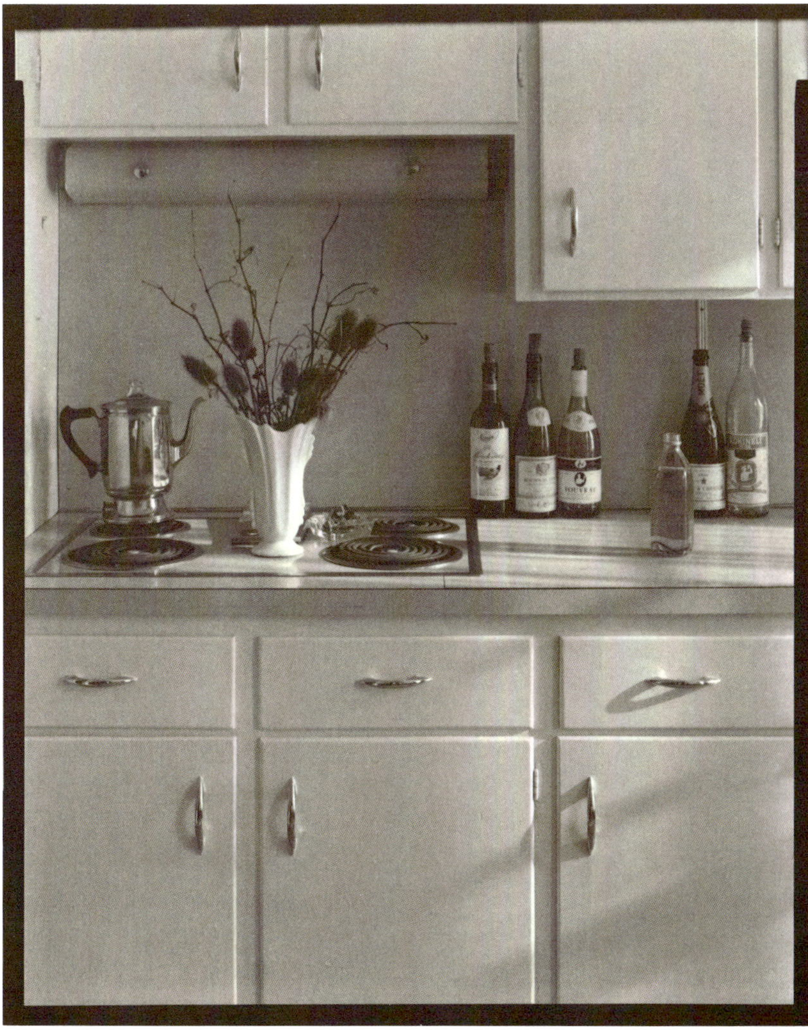

Figure 8. The finished Platinum Print.

24 Chemical Additives to the Standard Sensitizer Formula

There are some chemicals that when added to the standard sensitizer formula will effect a color or contrast change to the image. However, as all but one are considered highly toxic, the only one recommended for general use is Gold Chloride. It is usually made in a 2% solution and a few drops can be added directly to the A, B and C formula. Further details as to the effect of adding Gold Chloride to the sensitizer are included in the next section of this manual. It is highly recommended that you avoid working with Lead Acetate, Mercuric Chloride and Uranium Nitrate because of their known toxicity.

Increasing the Contrast of the Print

One difficulty faced by the silver printer moving over to Platinum printing is the significant reduction in contrast in the Platinum Print. This effect can be adjusted by following these suggestions:

1. Increase the amount of B solution in the sensitizer formula.

2. Double coat your paper. The results are worth the extra time and effort.

3. Increase the exposure time by moving the light source farther from the printing frame.

4. Varnish the paper surface after processing and drying. This will make the paper surface more reflective and enhance the contrast of the print. This procedure is questionable as the wax and varnish will probably yellow or darken with time.

5. Add about 15 drops of a 10% solution of Potassium Dichromate to the developer solution. This has the disadvantage that it will increase the contrast of all the prints processed in that developer solution. The Potassium Dichromate will eventually become exhausted and will have to be replenished. This technqiue will not work with Palladium as the Potassium Dichromate will bleach the Palladium image.

6. Mix a 1:12 ratio of Polymer Gloss medium in distilled water and brush on 5 cc of this solution onto a processed print.

7. Process your exposed print in either Sodium Citrate or Sodium Acetate as opposed to Potassium Oxalate. Because these developers are less active than Potassium Oxalate, they will require longer exposure times for your print.

25 Decreasing the Contrast of the Print

The following are methods used to reduce the contrast of a print:

1. Add a few drops of a 2% solution of Gold Chloride to the sensitizer formula. This will add a tone to the highlights of the print, thereby compressing the tonal range and reducing contrast.

2. Use less solution B in the sensitizer formula.

3. Reduce the exposure time.

4. Use a rougher surfaced paper support which will scatter the light reflected off the print surface.

5. Single coat your paper.

6. Palladium usually produces less sharpness and contrast than Platinum.

7. Print the paper while slightly damp. This will greatly suppress the contrast and may impart a lovely blue tint to the image.

8. Process your exposed print in Potassium Oxalate rather than Sodium Citrate or Sodium Acetate. Potassium Oxalate is a more vigorous developer than Sodium Acetate and Sodium Citrate, and will therefore require less exposure time for your negative.

To Produce Warm-Toned Prints

The following procedures will produce warm tones in the finished print:

1. Use only Palladium in the C solution of the sensitizer.

2. Add a few drops of a 2% solution of Gold Chloride to the sensitizer formula.

3. Process your print in Potassium Oxalate instead of Sodium Acetate or Sodium Citrate.

4. Use a warm-toned or cream color paper support.

26 To Produce Cold-Toned Prints

1. Use only Platinum in the C solution and process the print in either Sodium Acetate or Sodium Citrate.

2. Add Potassium Phosphate to the Potassium Oxalate developer in the following proportions:

   A. Water (distilled at about 90 Degrees F (32 degrees C) 1000ml
   B. Potassium Oxalate .... 180 grams
   C. Potassium Phosphate .... 60 grams

   Use the solution at room temperature.

3. Processing a Palladium print in Sodium Acetate or Sodium Citrate will produce an image color almost identical to a pure Platinum Print.

Possible Problems: Their Causes and Remedies

Because of the simplicity of the Platinotype Process, there are only a few possible problems that may develop during printing. These problems are usually easily defined and just as easily corrected. As stated earlier, the more conscientious and consistent you are in your work habits, the greater will be your immediate and continued success with this process. Following is a list of potential problems and their causes as well as suggestions for their correction. This list is not comprehensive and you may encounter problems unique to your working situation. This will call for a careful analysis of your working procedures, environment and materials. A simple and effective way of solving a unique problem is to begin to change one variable at a time until the problem's source is discovered: i.e., change your paper support, developer solution, A and B solution, water supply, etc.

Problem...Solution

A. Your finished prints appear vigorous in the shadows and mid-tone areas, but appear fogged or degraded in the highlights. Possible causes and recommended solutions...

1. The sensitized paper was exposed to UV radiation either during coating, drying, or inspection of the print during exposure or processing. Always protect the sensitized paper from UV radiation until the print is completely etched. This includes not viewing the print during processing by daylight or fluorescent light.

2. The A and B solutions may be outdated and the Ferric Oxalate may have decomposed into the ferrous state. Check to see if any precipitate has settled on the bottom of the bottles. A simple method of checking the state of the Ferric Oxalate is to place a small amount of the Ferric Oxalate into a glass container and add 1 drop of 5% solution to Potassium Ferricyanide (Farmer's Reducer). Shake well and then observe the density of the blue color that appears in the liquid. The deeper the blue color, the more Ferrous Oxalate present. The color change to blue should be subtle, not pronounced, and, of course, this test must be conducted under a weak incandescent light source.

3. The print was overexposed during printing. Simply reduce the exposure time for the next print.

B. Your prints do not develop up to full intensity.

1. The paper was printed while still damp. Increase your drying time and dry more thoroughly with the hair dryer. Remember to dry both the front and the back of the paper. Avoid printing in excessively humid environments.

2. The print was underexposed during printing. Increase your printing time. Making test prints to determine the correct exposure time will help eliminate this problem.

3. The developer solution may be exhausted. This is more likely to happen with Sodium Acetate and Sodium Citrate than with Potassium Oxalate. Decant off any sludge that may settle in the developer solution. If necessary, mix fresh developer. Do not use any of the developer solutions below 60 degrees F (16 degrres C).

4. The negative may not be of sufficient contrast for the Platinotype Process, or you did not use enough B solution to compensate for the low contrast of the negative.

C. The print appears grainy after processing.

This will occur if too much of the B solution is used in the sensitizer formula, or if the temperature of your developer is below 60 degrees F (16 degrees C). This effect may also occur when using a fresh developer solution. The graininess is eliminated after passing a few prints through the fresh developer.

D. The print looks good in the developer, but appears weak upon drying.

1. The sensitizer was allowed to sink too deeply into the paper fiber during drying. Reduce your drying time and use a cooler temperature setting on the hair dryer.

2. The developer may have become too strongly alkaline through continued use. It should be just slightly acidic. This can be checked with the use of litmus paper.

E. The finished print is too strong in contrast.

1. This is usually caused by too much B solution in the sensitizer formula.

2. The print was underexposed. Increase your exposure time on the next print.

F. Spots and streaks in the print.

1. Working with contaminated or dirty materials, or touching the surface of the print with your hands.

2. Air bubbles formed on the surface of the print during development. These can be eliminated by being brushed over with a styrofoam brush as the print is being developed.

3. Impurities in the paper fiber. Check each sheet of paper prior to sensitizing, via transmitted light.

4. Dark spots with a comet-like tail will appear on the print from any metal salts that have not gone into solution. Re-heat and remix the C solution prior to using.

5. Black specks in the print may be caused by processing in a developed solution that has accumulated particles of metal from extended usage. Switch to a fresh developer solution.

6. Mottled or blotchy developed prints. Some paper sizing materials will react with the sensitizer solutions to produce these results. Switch to a different paper stock.

7. Blue or lavender tones in the highlight areas of the print. This is caused by adding too much Gold Chloride to Sensitizer solution. Reduce the number of drops of the Gold Chloride, 3 or 4 drops in a sensitizer for an 8 × 10 inch negative is sufficient.

29

30 Bibliography

Monographs, Manuals and Treatise on the Platinum Process

Abney, Captain, Platinotype, Its Preparation and Manipulation, London, Sampson, Low, Marston Company, 1895.

Anderson, Paul L, *Pictorial Photography: Its Principles and Practice,* Philadelphia, H.B. Lippincott, 1917.

Arno, Jan, *Handbook of Alternative Photographic Processes*, New York, Van Nostrand Reinhold Company, 1982.

Bunnell, Peter C, *Non-Silver Printing Process*, New York, Arno Press, 1973.

Burbank, W.H., *Photographic Printing Methods*, New York, The Scovil & Adams Company, 1891.

Burnett, C.J., *On the Production of Direct Positives on Printing by the Salts of the Uranic and Ferric Oxides.* London, The Photographic Fine Arts Journal, Vol. XI, No. VII. June 1858.

Crawford, William, *The Keepers of Light*, New York, Morgan and Morgan, 1979.

Dawson, George, *The Platinotype Printing Process*, London, British Journal of Photography, Vol. 27, No. 1029, 1880.

Hafey, John and Shillea, Tom, *The Platinum Print*, Rochester, The Graphic Arts Research Center, 1979.

Hinton, Horsley A., *Printing, a Simple Book on the Process.* London, Watson and Vinety, Ltd., 1897.

Lietz, Ernst, *Modern Heliographic Processes*, New York, D. Nostrand Company, 1888.

Nadeau, Louis, *The History and Practice of Platinum Printing*, Fredericton, New Brunswick, 1984.

Pizzighelli, Captain and Hubl, Arthur Baron V., *Platinotype*, London, Harrison and Harrison, 1866.

Rexroth, Nancy, *The Platinotype*, 1977. Yellow Springs, Violet Press, 1977.

Warren, W.J., *The Platinotype Process of Photographic Printing*, London, Iliffe Sons and Strumey Lts., N.D.

Willis, William, *The Platinotype Process*, London, British Journal of Photography, Vol. 27, No. 1028, 1880.

General References

Adams, Ansel, *The Negative*, Boston, The New York Graphic Society, 1981.

Bureau, William H., *What the Printer Should Know About Paper*, Pittsburgh, Graphic Arts Technical Foundation, 1982.

Carroll, John S., *Photographic Facts and Formulas*, New Jersey, Prentice-Hall, Inc., 1976.

Eaton, George T., *Photographic Chemistry, 2nd Edition.* Hastings-on-Hudson, Morgan and Morgan, 1965.

Hunter, Dard, *Papermaking, the History and Technique of an Ancient Craft*, 2nd Edition, New York, Alfred A. Knopf, 1967.

Jones, Bernard E., *Encyclopedia of Photography*, New York, Arno Press, 1974.

Keefe, Laurence W. and Inch, Denis, *The Life of a Photograph*, London and New York, Focal Press, 1964.

Kodak Publication, *The Preservation of Photographs*, New York, Eastman Kodak Company, Publication No. F-30, 1979.

Kodak Publication, *Copying and Duplicating in Black-and-White and Color*, New York, Eastman Kodak Company. Publication No. M-1, 1984.

Kosar, Jaromir, *Light Sensitive Systems: Chemistry and Application of Non-Silver/Halide Photographic Processes*, New York, John Wiley and Sons, 1965.

Todd, Hollis N., *Photographic Sensitometry, A Self-Teaching Text*, New York, John Wiley and Sons, 1976.

Appendix A

Step-by-Step Outline for Making a Platinum Print

1. Assemble the A, B and C solutions along with your glassware and sensitizing brush.

2. Fill tray with developer solution and mix the three acid-bath solutions.

3. Set up the exposing light source. If using a sun-lamp, turn on the lamp.

4. Select the sensitizer formula suited for the negative to be printed.

5. Select the paper to be sensitized and inspect it for impurities and imperfections.

6. Tape the paper to a clean sheet of glass or plexiglass.

7. Measure out the drops of the sensitizer formula. Wet the brush and squeeze dry.

8. Pour the sensitizer solution in the center of the paper and evenly brush the solution over the inscribed area.

9. Air dry the paper for 10-12 minutes and then blow-dry for 1-2 minutes with the hairdryer set on low or medium heat.

10. Assemble the sensitized paper, negative and contact frame.

11. Place the assembled contact frame under the exposing light source. Begin the exposure and make a notation of the time.

12. As soon as the unprotected edges of the sensitized paper begin to darken, remove the contact frame from the light source, open half of the hinged back of the frame and visually inspect the provisional image.

13. Continue exposing the print until the shadow details are clearly visible and the midtones are slightly visible.

14. Remove the exposed print from the contact frame, place it face-up in a developer solution, and brush over any air bubbles that may form. Develop the print for at least two minutes.

15. Etch the developed print, with monitoring, in the three acid-baths. Etching time is 5 minutes in each bath.

16. Wash the etched print in running tap water for approximately 30 to 60 minutes.

17. Hang the washed print up in the darkroom to airdry. If necessary, flatten the airdried print in a drymount press set at a temperature of 150 degrees F.

18. If required, spot the print using Kodak or Spot-Tone dyes.

19. Photo-corner the finished print onto a piece of acid-free conservation board, and if desired, overmat the print. Use only acid-free materials for all finishing procedures.

Appendix B

Black-and-White Film Requirements

1. Making Enlarged Duplicate Negatives

The Platinotype Process is a contact printing process. This means that the image size will be the same as the negative size, the negative being printed in "contact" with the sensitized Platinum paper. The ideal method of obtaining your contact print negative is to make your photograph with a view camera. However, if this is not possible or desirable, an enlarged black and white negative can be made from a miniature negative in a simple one-step procedure. The standard method is to use Kodak's Black-and-White Duplicating Film #4168. It is available in 4 × 5, 5 × 7 and 8 × 10 sheets.

This film is a blue-sensitive material, which means it can be handled under a standard No. 01 Red Safelight. The film has a transparent base which makes identifying the emulsion side under illumination a bit difficult. If you hold a sheet of film in your hand in the vertical position and place the notch of the film in the upper right hand corner, then the emulsion side of the film will be facing you.

The #4168 film has already been exposed by the manufacturer, which means it will work like a color transparency film. More exposure produces less density (lighter results) and less exposure produces more density (darker results). Simply remember: "less is more" in terms of exposure.

When making your enlarged duplicate negative, handle the #4168 film as if it were a black-and-white paper. The original negative is placed in the enlarger, the image magnification is selected, and the image is focused on the easel. Place a piece of black mat paper on the easel under the film to reduce the chance of halation (light reflecting back through the film during exposure) and lay a sheet of the #4168 film, emulsion side up, on the easel.

To determine your correct exposure time, make a test-strip exposure as you would with photographic paper. Use an aperture of F/4 on the enlarger lens and start with a 20 second exposure, and continue with exposures at 5 second intervals, up to 60 seconds. Process the negative in a tray as you would a print. Use Kodak Dektol Developer diluted to 1:10 at a temperature of 70 degrees F (21 degrees C), for about 3½ minutes with constant agitation. Development has a "normal" effect on this film. Longer development times increase contrast and shorter development times decrease contrast. Therefore, density is controlled by exposure time and contrast is controlled by developer concentration and development time.

After development, fixation and washing, inspect your duplicate negative and select the exposure time that looks "good". Use this exposure time to make your full image. As with any photographic material, some practice will be necessary to appreciate the full potential of this film. Keeping exact notations and using consistent working procedures will guarantee superior results.

One possible disadvantage of this film is the high base + fog density which is produced after exposure and development. This additional density may require prolonged exposure time for your print. A good way of controlling this situation is to cut a small strip of the #4168 film from a larger sheet and expose it to white light in your darkroom for at least one minute and then process the strip as you would a normal sheet of the film. This greatly overexposed film strip will indicate the clearest base + fog possible with this film. Try to match the base + fog density on your duplicate negative to this test strip.[1]

Another procedure for reducing the excess base + fog density of this film is to remove some of the silver forming the image with a chemical called Farmers Reducer (Potassium Ferricyanide), which is available at any photo supply store. The procedure is as follows:

Mix about a half-cap full (using the cap from the Potassium Ferricyanide bottle) of the Potassium Ferricyanide in 50 ounces (1478 ml) of tap water at approximately 70 degrees F (21 degrees C). Soak the processed #4168 negative in this solution for approximately 2 minutes. The Potassium Ferricyanide will soften the film's emulsion, so handle the film gently during this procedure. Then simply re-fix, re-wash and dry the film. This procedure should only be necessary if you cannot control the density of the negative through proper exposure.

A second, but more involved method of making a duplicate enlarged negative is to use the interpositive or diapositive technique. A clear and complete explanation of this procedure is contained in the Kodak Manual, entitled ***Copying and Duplication in Black-and-White and Color***, Code #M-1, page 123.

A third method of making an enlarged duplicate negative is to use a 35mm color slide. Project the slide via an enlarger, as you would a miniature black-and-white negative, onto a piece of panchromatic sheet film. Try working with Kodak Plus-X or Super XX film for this purpose. As these films are more light sensitive than the #4168 film, your exposure time will be much shorter with the use of a smaller aperture on your enlarger lens. Try your initial test-strip exposure at F/16 for 3 seconds and continue at 2 second intervals to a total of 15 seconds. Process the

1. Crawford, William, *The Keepers of Light*, New York, Morgan and Morgan, 1979. Page 125.

panchromatic film in Kodak HC-110 developer diluter 1:7 (this is the B dilution on the bottle). Use the development temperature and time recommended for the B dilution. Because these films are panchromatic, they must be handled and processed in total darkness.

A fourth method of acquiring an enlarged black-and-white negative suitable for platinum printing is to make a black-and-white print, dodged and burned to meet your specifications. Then have a copy negative made from your original print at a professional photo lab. The fee for this service is usually quite reasonable and produces a satisfactory negative.

2. Black-and-White Film Characteristics

Because the Platinotype Process allows for substantial manipulation during the sensitizing stage, it is possible to use a broad range of contrast and density negatives in the process. A soft negative, one with little density or contrast, can product either a soft or hard contrast Platinum Print. However, it is always preferable to design your negative's density and contrast to suit your aesthetic needs as represented by the finished Platinum Print. Some basic concepts concerning negative/positive relationship are:

1. A thin "flat" negative will produce a very soft, "painterly" platinum image. It will, however, require a more contrasty sensitizer formula than normal and may result in extended printing time.

2. A "normal" negative, suitable for printing on a grade 2 or 3 silver paper will require the standard sensitizer formula and an average exposure time, i.e. 15-20 minutes under two 275 watt sunlamps.

3. A dense and contrasty negative, suitable for printing on a grade 0 or 1 silver paper, will require a less contrasty sensitizer formula and therefore, a shortened exposure time, i.e. 8-10 minutes under two 275 watt sunlamps.

It will, of course, be more difficult to produce a "soft" quality print from a dense and contrasty negative and vice versa. It is, therefore, to the photographer's advantage to previsualize his subject at the moment of exposure and to follow through with this previsualization in development to produce the negative best suited to make the desired print.

The author has designed his negatives to produce a density range of approximately 1.20 to 1.30, suitable for printing on grades 0 or 1 silver papers. The average printing times for these negatives is 10-12 minutes, and the resulting prints contain a full tonal range with the look and "feel" of the classic platinum prints from the turn-of-the-century.

Generally speaking, therefore, it is advisable to produce a negative with strong density and contrast for use in the Platinotype Process. This can be achieved by giving ample exposure to your film and then overdeveloping about 15-30% more than normal.

Appendix C

Paper Characteristics

The basic ingredient in the manufacture of paper is cellulose. Most modern papers are manufactured from wood pulp, which contains only about 50% cellulose. The remainder of the pulp is made up of about 30% lignin and small amounts of other carbohydrates and proteins. Unfortunately, for any papers intended for archival permanence or fine-quality work, the lignin contained in the pulp will gradually break down into acid by-products which will eventually disintegrate the paper fiber. Therefore, all the lignin must be removed from the highest quality papers.

Traditionally, the best papers have been made from 100% rag or cotton fibers. Sources for the cotton fibers are pure cotton and waste cuttings and threads from textile manufacturing. Cotton fibers are over 98% cellulose, and they are longer than wood fibers, as well as naturally intertwined. These characteristics impart a high tensile strength to the cotton fibers which make them an ideal substance for the manufacture of the highest quality papers.

1. Surface Texture

Most papers today are machine made. The most common type of machine is the Fourdrinier Screen, which utilizes a continuous belt made of a fine metal screen on which the saturated paper pulp is spread. As the screened belt moves through the machine, the water in the pulp is drained off and the paper fibers are allowed to interlock to form the basic paper stock. The top of the paper being formed on the screen is covered by a felt blanket. When the paper has dried, the texture of the screen can be seen; this is called the Screen Side, or Wire Side. The opposite side is called the Felt Side. A recent innovation in machine made paper is to use a screen on both the top and bottom of the paper as it is formed. This type of paper would, of course, have identical textured surfaces. The Felt Side is usually the preferred side to use when hand-coating a paper, as it is smoother than the Screen Side. Sometimes, it is difficult to see the difference between the Screen Side and the Felt Side until the paper has been moistened. Two simple methods of determining the Felt Side of a paper are:

1. The side on which the watermark reads correctly is the Felt Side.

2. Paper supplied in pads has the Felt Side facing up.

The other type of machine made paper is called Mold-Made paper. Mold-Made papers have the look and feel of handmade papers, including deckled (rough) edges.

This machine produces paper in individual sheets as opposed to a continuous roll (web) as on the Fourdrinier Machine. Mold-Made sheets also have a Screen and Felt Side.

Additional surface manipulation is carried out after manufacturing. This usually involves pressing the paper between smooth metal rollers. If the rollers are hot, the paper becomes Supercalendered, or hot pressed, which imparts a very smooth, shiny surface to the paper. If the metal rollers are cold, then the paper is classified as cold pressed, and will have a rougher, more matte, surface.

2. Grain Direction

Another consequence of paper being manufactured on a Fourdrinier Screen Machine is that fibers tend to line up in the direction that the screen passes through the machine during manufacturing. This forms a grain direction in the paper. This grain direction can be of consequence if you intend to multiple print in registration on the sheet of paper. Paper will expand and contract more in the cross-grained direction than in the grain direction. This problem is eliminated with Mold-Made papers, as they tend to expand and contract evenly in both directions, due to the random formation of the fibers during manufacturing. The Grain Direction of a piece of paper can be determined by floating a square piece of the paper on water, or simply by wetting one side of the paper. The paper will curl away from the moistened side, and the axis of the curl will be parallel to its grain direction. When intending to multiple print on a piece of paper, it is best to place the negative on the paper with its long side parallel to the grain direction of the paper, as this is the direction of least physical change during printing.

3. Sizing

The purpose of sizing paper is to partially seal the space between the fibers to make the paper more impervious to water and solutions. Without some type of sizing, paper would be like blotting paper and absorb all liquid applied to it. Papers that are unsized and absorb water rapidly are called Waterleaf. Papers that have a low degree of sizing are called slack-sized or quarter-sized, and papers that have a high degree of sizing are called hard-sized or full-sized.

There are two basic methods used in sizing commercially manufactured papers. They are internal sizing and external sizing. The purpose of internal sizing is to retard the penetration of water or other fluids into the paper, but it does not make the paper waterproof. [2] The most common method of internal sizing is the Alum-Rosin Sizing technique. Unfortunately, the Alum size imparts a distinct acid state, usually in the pH range of 5.0.

External sizing, or surface sizing, differs from internal sizing in its application and purpose. Surface sizing usually consists of applying heated starch solution to the partially dried web at the paper machine press. Sizing is forced into both sides of the paper and penetrates below the surface to some extent. It normally contributes little water resistance. One of its purposes is to cement, or seal down, surface fibers and thus increase the surface strength of the paper.[3]

Additives like Melamine-Formaldehyde and Urea-Formaldehyde resins are sometimes used as external sizing for papers. These papers will emit a strong scent of formaldehyde when being coated with the Platinum Sensitizer, and may produce a streaked and grainy print. Papers sized with these resins are not highly recommended for platinum printing. Because all machine papers are either internally or externally sized by the manufacturer, it is not necessary for the Platinum Printer to hand-size his paper before applying the sensitizer. However, if the worker is still interested in hand-sizing his paper, formulas for the classic sizings are contained in the bibliographic references.

4. The Issue of Permanence

Because the noble metals Platinum and Palladium are quite stable, they form very permanent photographic images. The weak link in the chain of permanence, besides improper processing and print washing, is the quality of the paper support on which the image is printed. The single most critical factor in the deterioration of the paper is the presence of acid. The best papers for archival permanence are papers made of 100% cotton or linen, sized without alum, in an acid-free condition, with a pH rating between 6.5 and 7.5. Another precaution taken by some paper manufacturers is to add a "buffer" to the paper, usually Calcium Carbonate.

There are many fine quality 100% rag papers on the commercial market. Some papers that have been traditionally used for the Platinotype Process are: Strathmore Bristol, Series 500; Fabriano 901 lb. hot press; Arches, D'Arches watercolor paper; Rieves BFK and Rieves Lightweight, and more recently, Rising Artist Drawing Bristol, plate finish.

However, just because a paper is classified as 100% rag does not necessarily mean that it is acid-free or buffered. The serious photographer will experiment with a range of papers and investigate their pH levels, and in this empirical way, select a few types of paper to suit his needs.

Information on the method used to manufacture the paper, types of sizing used in the paper, its pH level and whether or not it is buffered, is usually available through the retail paper suppliers or directly from the wholesale distributors or manufacturer.

As with any commercially manufactured product, even the highest quality papers will vary from batch to batch in their printing characteristics. It is not uncommon to detect a definite difference in the way a particular brand of paper will perform over an extended period of time. Also, some papers work better with Platinum or Palladium, or work only with one metal salt and not with the other. The best way to determine this characteristic of a paper is to make a small test print with the desired metal salt before attempting to make a full-size print.

5. Depositing an Alkaline Reserve

a. Neutralization of Acidic Residues*

Residual acidity cannot always be diminished in washing cycles, and can eventually lead to possible embrittlement and deterioration of the print. De-acidification is the practice of placing an alkaline reserve in the paper fibers as a buffer against further acid-induced deterioration. In the conservation of paper, many methods have been developed to accomplish this and to this date, no one method is optimal. In addition, the long-term effects of placing an alkaline buffer in photographic materials are not fully known, so it is advisable to avoid de-acidification processes until further testing is completed.

The alternative which neutralizes inherent acidity in the paper is immersion of the print in a dilute alkaline bath. An alkaline rinse is a distilled water bath to which calcium hydroxide has been added to create a mildly alkaline bath (pH 8.5-9.0). The procedure is described below.

*This information supplied by Ms Holly Maxson.

b. Preparing a Calcium Hydroxide Aqueous Solution

Begin with one litre (approximately one quart) of distilled water in a clean plastic or glass container. Slowly add 2 grams of Calcium Oxide powder while stirring vigorously. This proportion results in a saturated solution and any excess powder precipitates out and settles on the bottom of the container. The pH of this concentrated stock solution should be around 12.4. Keep the container tightly closed when not in use.

To prepare a working solution from this stock concentrate, decant one ounce of the clear solution without disturbing the precipitate on the bottom of the container, and mix it with one gallon of distilled water. If possible, the pH should be no more alkaline than 8.5. This condition can be easily checked with the use of a standard pH paper. When the clear stock solution is depleted, more water can be added, and the deposit stirred up to create another saturated solution. This process can be repeated (adding water without additional Calcium Oxide powder) 5 or 6 times before the container should be washed out and a new concentrate made with fresh powder.

c. Use of the Alkaline Rinse

The dilute alkaline bath (pH 8.5) is made as described above, and the washed print immersed in the solution for approximately one half hour. When the print is removed from the bath, it is allowed to air dry. During the drying process, the Calcium Hydroxide combines with the Carbon Dioxide in the air, leaving a fine deposit of Calcium Carbonate in the paper fibers.

The alkaline rinse should effectively neutralize all acidic groups in the paper, although it can provide no protection against future acid-induced deterioration. Of course, this alkaline rinse procedure is conducted after the thorough water washing of the print, as described on pages 23 and 24 of this manual. The print should not be water washed again after soaking in the alkaline bath, as this would wash out the alkaline reserve deposited in the paper fibers.

*This information supplied by Ms Holly Maxson.

2. Bureau, William H., *What the Printer Should Know About Paper.* Pittsburgh, Graphic Arts Technical Foundation, 1988, Page 60.
3. Ibid., Page 61.

Appendix D

Chemistry of the Platinotype Process

Six chemical compounds are needed to make a platinum print:

1. Potassium Chloroplatinite or Sodium Chloropalladite
2. Ferric Oxalate
3. Oxalic Acid
4. Potassium Chlorate
5. Citric Acid*
6. Potassium Oxalate†

*In the original process, Hydrochloric Acid was used instead of Citric Acid. However, as HC1 is a highly corrosive chemical, it is recommended that you use Citric Acid, which is non-corrosive and non-toxic.

†Potassium Oxalate is the classic developer used in the Platinotype process. Other chemical compounds often used as substitutes for Potassium Oxalate are Sodium Citrate and Sodium Acetate.

1. Chemical Formulas

The classic formula for mixing the solutions used in the Platin… ss are given below.

1. Solution A
   - A. Distilled water .... 6(
   - B. Oxalic Acid .... 1.1 g
   - C. Ferric Oxalate .... 16 grai

2. Solution B
   - A. Distilled water .... 60 ml
   - B. Oxalic Acid .... 1.1 gram
   - C. Ferric Oxalate .... 16 grams
   - D. Potassium Chlorate .... 0.30 grams

3. Solution C (Platinum or Palladium metal salts)
   - A. Distilled water .... 60 ml
   - B. Potassium Chloroplatinite
     or
     Sodium Chloropalladite .... 10 grams

Smaller quantities can be prepared as long as the proper proportions are maintained. The ratio of Platinum or Palladium Metal Salt to water is 1 gram to 6 mls. The A and B solutions should be mixed at a temperature of 80 degrees F to 90 degrees F (27 to 32 degrees C).

The C solution(s) should be mixed at a temperature of approximately 115 degrees F (52 C). Some of the metal salts in Solution C may come out of solution during prolonged storage. If this occurs, gently heat the solution and the salt will redissolve.

The shelf-life of the A and B solutions is approximately 6 to 8 months, depending on storage conditions. The shelf-life of the C solutions is indefinite due to the extreme stability of the Platinum and Palladium Metals.

Appendix E

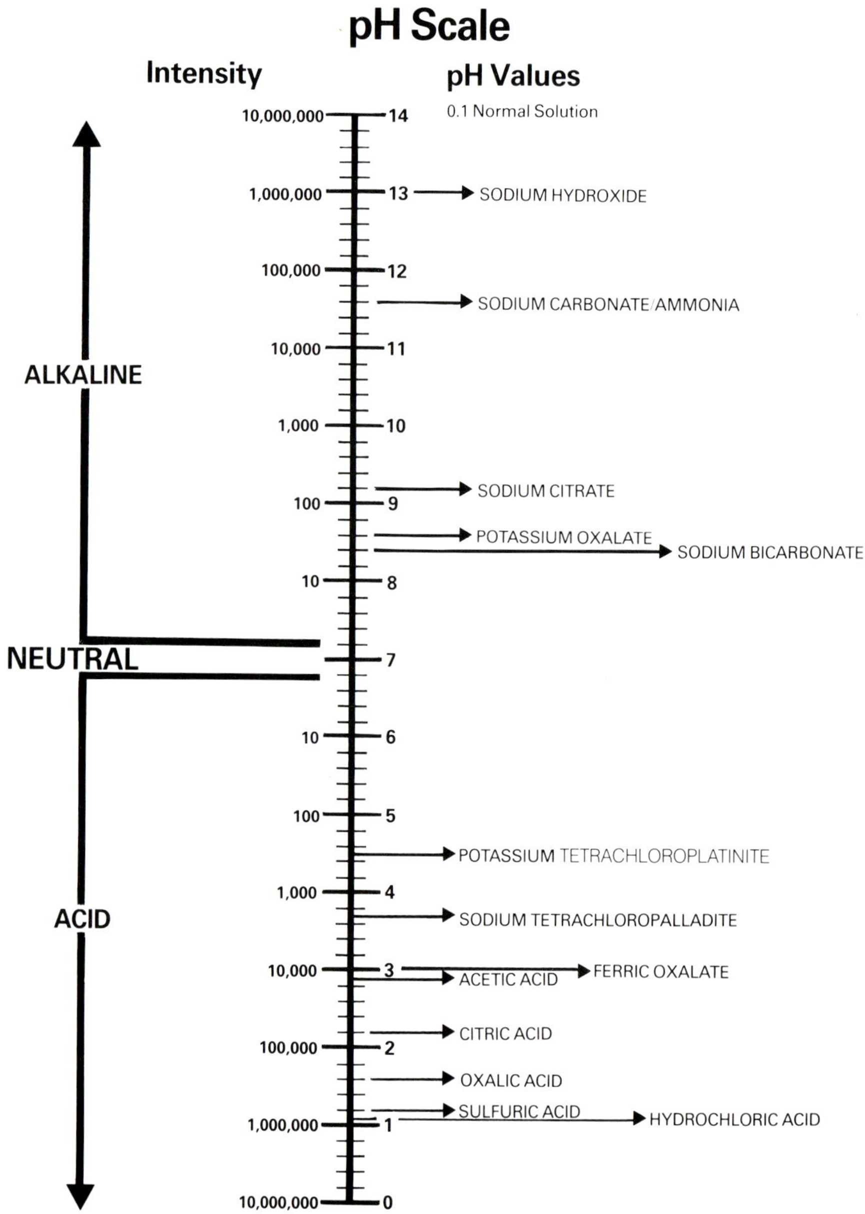

A change of one number on the pH scale represents a ten-fold change of intensity of the acidity or alkalinity of the solution.

C Table

Avoirdup t

1 dram = 27.

1 ounce = 16 dr 5 grains

16 ounces = 1 pou 5 grams

1000 grams = 1 kilog

Example: Determine the ounds in one kilogram of Potassium Oxalate:

1 kilogram × 1000 grams / 1 kilogram × ... / 4 .6 gms = 2.2 pounds

Fluid Measures

1 ounce = 29.5 milliliters

32 ounces = 1 quart = 0.95 liters

1 gallon = 4 quarts = 3.78 liters

The above represent U.S. Fluid Measures. The British or Imperial quart contains 40 fluid ounces. As in the U.S. system, the Imperial quart contains two pints and the gallon contains four quarts.

Example: Determine the number of milliliters contained in one quart of water:

$$1 \text{ quart} \times \frac{32 \text{ ounces}}{1 \text{ quart}} \times \frac{29.5 \text{ milliliters}}{1 \text{ ounce}} = 944 \text{ milliliters}$$

Temperature

There are two common temperature scales, Centigrade and Fahrenheit. In the Centigrade scale, denoted as C, the freezing point of water is zero degrees C, and the boiling point of water is 100 degrees C. Thus, this scale has 100 degrees between the two standard reference points. The Centigrade scale is used in all scientific work.

In the Fahrenheit scale, denoted F, the freezing point of water is 32 degrees F, and the boiling point of water is 212 degrees F. Thus, in this scale, there are 180 degrees between the two standard reference points. The two temperature scales are related by the following expressions:

$(C \times 9/5) + 32 = \text{degrees F}$ $(F - 32) \times 5/9 = \text{degrees C}$

For example, to convert a temperature of 20 degrees C to Fahrenheit, one would...

20 degrees C $\times$ $9/5$ or $20 \times 1.8 = 36 + 32 = 68$ degrees F

Likewise, to convert a temperature of 68 degrees F to C, one would…

68 degrees F $- 32 = 36$ and $36 \times 5/9$ or $36 \times 0.55 = 19.8$
(rounded off to 20 degrees C).

Some Common Prefixes

Centi: 0.01 (one hundredth)
Milli: 0.001 (one thousandth)
Kilo: 1000 (one thousand times)

Weight Percentage

The weight percentage of a compound in a solution is 100 times the weight of compound divided by the total weight of solution. Thus, 100 grams of a 10% aqueous solution of Citric Acid contains 10 grams of Citric Acid and 90 grams of water.

$$\% = 100 \times \frac{\text{Weight of Compound}}{\text{Total Weight of Solution}}$$

Example: What weight of water and Sodium Citrate should be used to prepare 300 grams of a 30% Aqueous Sodium Citrate solution?

$$\text{Grams of Sodium Citrate} = \frac{(\%) \times (\text{Total Weight of Solution}}{100}$$

$$= \frac{(30) \times (300)}{100} = 90 \text{ grams}$$

Grams of water = 300 grams − 90 grams = 210 grams

47 Appendix F

Safe Handling of Chemicals

1. General Information

All chemicals, if mishandled, are dangerous. Therefore, when working with the chemicals used in making the Platinum and Palladium prints, the worker must handle these chemicals with intelligence and care. In most cases, no harmful effects need be feared if precautions are taken to avoid the following:

1. Skin contact
2. Inhalation of chemical dust or fumes
3. Ingestion

While many of the chemicals used in the Platinotype process may be fatal if ingested, they should not present any hazard providing they are properly packaged and sealed in safety containers; labeled and stored away from children and pets. Common sense should tell one that these chemicals, as well as many household chemicals, demand respect as to their handling, packaging, labeling and storage.

Some individuals may develop an allergic reaction to the chemicals used in the Platinotype process which would of course preclude their working with these chemicals. Any such allergic reaction will be eliminated by removal of the cause of the irritation.

Two excellent and comprehensive sources of information on the physiological effects of the chemicals listed in this manual are:

1. **Overexposure: Health Hazards in Photography**, listed in the bibliography of this manual.

2. **Dangerous Properties of Industrial Materials**, by N. Irving Sax, New York, Van Nostrand Reinhold Co., 1975.

2. Safety Data

1. Citric Acid $C_6H_8O_7 \cdot H_2O$
   Concentrated Citric Acid is moderately toxic by skin contact, inhalation and ingestion. Dilute solutions are mildly irritating and may cause allergies in some individuals. Citric Acid may be substituted for Hydrochloric Acid as a clearing bath for platinum prints.

2. Potassium Chloroplatinite K_2PtCl_4
   Repeated exposure to Platinum salts in powder form can cause skin and respiratory allergies in some individuals. Always wear gloves and safety goggles when handling the Platinum salts. Mix in a fume hood or glove box or wear an approved toxic dust mask when mixing the powder into solution.

3. Ferric Oxalate $Fe_2(C_2O_4)_3$
   Ferric Oxalate is a mild irritant which may be fatal if acutely ingested. Oxalates are corrosive to the skin and eyes, nose and throat. Avoid continuous skin contact.

4. Gold Chloride $HAuCl_4$
   Gold Chloride is moderately toxic by every route of exposure. It can cause severe skin and respiratory allergies. Possible irreversible damage can occur if inhaled or ingested. Avoid skin contact.

5. Hydrochloric Acid HCl
   Hydrochloric Acid is used more commonly than Citric Acid as a clearing bath for Platinum Prints. Inhalation or ingestion of concentrated amounts may be fatal. Small concentrations of the fumes will irritate mucous membranes, eyes and respiratory tract. Hydrochloric Acid should be used only in a well ventilated area with adequate personal protection in case of accidental spillage or splattering. Avoid skin contact and inhalation of fumes.

6. Mercuric Chloride $HgCl_2$
   Mercuric Chloride is not recommended by this author for general usage. This compound is an extremely toxic and highly absorbable irritant and allergin. Ingestion, inhalation or absorption of small quantities can be fatal. Handle with extreme caution.

7. Oxalic Acid $C_2H_2O_4 \cdot H_2O$
   Oxalic Acid is used in the production of Ferric Oxalate, as well as acting as a preservative for that solution. Oxalic Acid is a severe irritant and corrosive. Acute ingestion or inhalation may be fatal. Avoid skin contact.

8. Potassium Chlorate $KClO_3$
   Potassium Chlorate is moderately toxic by skin contact and ingestion, but may be highly toxic by inhalation. Combining potassium chlorate with hydrochloric acid will cause the release of highly toxic chlorine gas. Potassium chlorate is a fire hazard when in contact with flammable materials.

9. Potassium Oxalate $K_2C_2O_4 \cdot H_2O$
   Potassium Oxalate is moderately toxic by skin contact and highly toxic by inhalation and by ingestion. Ingestion of Potassium Oxalate may be fatal. Avoid skin contact and inhalation of its dust.

10. Sodium Chloropalladite Na_2PdCl_4
    Sodium Chloropalladite is slightly toxic by skin contact and by inhalation. Its hazards by ingestion are unknown, although it has been reported to cause internal problems to laboratory animals which have been administered large intravenous injections.

11. Sodium Acetate $NaC_2H_3O_2 \cdot H_2O$
    Sodium Acetate is slightly toxic by skin contact, inhalation and ingestion. Concentrated forms may be mildly irritating. Acute ingestion can cause purging.

In order to avoid the potential hazards of personally mixing the chemical compounds necessary to make the A, B and C solutions of the Platinotype process, it is possible to purchase the chemicals already mixed in solution. Following are a list of the major suppliers of all chemical solutions necessary to make a platinum print.

1. Blue Mountain Photo Supply
   P.O. Box 3085
   Reading, PA 19604

2. Elegant Images
   3328 N. N.W.
   Washington, DC 20007

3. Photographers Formulary
   P.O. Box 5105
   Missoula, MT 59806

4. Bostick and Sullivan
   P.O. Box 2155
   Van Nuys, CA 91404

5. Johnson Matthey
   2001 Nolte Drive
   West Deptford, NJ 08066

Appendix G

Glossary of Important Terms

Archival
Generally speaking, the preservation of important materials and documents. Photographically speaking, the proper chemical processing, handling, and storage of photographic materials to insure maximum stability and preservation.

Base + Fog Density
A minimum density (deposit of silver metal) on the unexposed edge of a chemically developed negative or print.

Buffer
A substance which, when added to a solution, causes a resistance in any change in pH.

Citric Acid
Chemical Formula: HO2CCH2C(OH) (CO2H)CH2(CO2H) Organic acid commonly found in foods, especially citrus fruit. A dilute solution of this acid is used to etch platinum prints.

Densitometer
A machine used in the science of sensitometry to measure the densities of exposed and processed emulsions. It is the densitometer that enables a photographer to take highlight and shadow readings from processed negatives to determine the density range of the negative.

Density Range
A numerical measurement of the difference between the highlight density and shadow density of an exposed and processed film. The density range is determined by subtracting the lesser density from the greater density. This measured density range allows a photographer to select the correct paper grade or, in the case of platinum printing, the correct sensitizer formula with which to print the negative.

Developing-out Paper
A modern type of photographic paper which produces a latent (invisible) image upon exposure to light, and must be chemically developed to produce the visible image.

Diapositive (Interpositive)
An intermediate positive image on film, made from an original negative, which is then used to make the finished copy negative. (See Kodak's Booklet M-1, *Copying and Duplicating in Black-and-White and Color,* p. 123.

Duplicate Negative
A copy or second-generation negative produced from an original negative or print.

Emulsion
A suspension of light sensitive silver halide crystals in a gelatin solution, usually bonded to a support or base, such as paper or plastic.

Ferric Oxalate
Chemical Formula: $Fe_2(C_2O_4)_3$ The Photosensitive chemical present in the platinotype sensitizer solutions. When paper coated with Ferric Oxalate is exposed to UV radiation, the iron is reduced from Fe(III) to Fe(II) and a provisional image is formed.

Gold Chloride
Chemical Formula: $AuC1_3$ or $HAuC1_4$. A few drops of a 2% solution of this compound, when added to the sensitizer, will deepen the color and reduce graining in Platinum Prints.

Litmus Paper
A type of paper which has been chemically treated with dye extracted from lichens rendering it sensitive to pH. Blue litmus paper turns red in the presence of acid. Red litmus paper turns blue in the presence of alkali.

Orthochromatic
An emulsion that is not sensitve to the red end of the visible spectrum. This type of emulsion can be handled under red safelight illumination.

Oxalic Acid
Chemical Formula: $H_2C_2O_4$. A naturally occuring organic acid present in many plants and vegetables. Oxalic Acid is a component of the sensitizer solutions used in Platinum Printing.

Panchromatic
An emulsion that is sensitive to the full range of the visible spectrum. This type of emulsion must be handled in total darkness.

pH
(Hydrogen Ion Concentration) A measure of the effective acidity or alkalinity of a solution. The pH scale runs from 0 to 14 with 7 being neutral. Acid solutions have a pH value less than 7 and alkaline solutions have a pH value greater than 7.

Photographic Density
The measurable amount of silver metal produced by the combination of exposure to light and chemical development of a photographic emulsion. The greater the amount of exposure to light and the longer the development time an emulsion receives, the greater the resultant density.

Platinum Metal Group
The group of metals comprised of Iridium, Osmium, Platinum, Palladium Rhodium and Ruthenium. Of these, both Platinum and Palladium are used in photographic printing.

Potassium Chlorate
Chemical Formula: KC103. An oxidizing agent which is present in the sensitizing solution B and is responsible for increasing the contrast in prints.

Potassium Chloroplatinite
Chemical Formula: K2PtC14. The compound most commonly used as the source of platinum for Platinum Printing.

Potassium Oxalate
Chemical Formula: K2C204. The potassium salt of Oxalic Acid. The most commonly used chemical for developing Platinum Prints.

Precipitate
A substance, usually a solid, which separates from a solution as a result of some physical or chemical change.

Previsualize
Mentally visualizing the finished print while studying the subject.

Print Contrast
The measurable range of difference between the darkest and lightest tone in a photographic print. The greater the measurable difference, the greater the contrast of the print.

Printing-Out Paper
An antique type of photographic paper, which produced the finished, visible image solely by exposure to light, without requiring chemical development.

Provisional Image
The image formed by exposure to UV radiation on a piece of platinum paper. The provisional image is used to determine, by visual judgment, the correct exposure time for the print. The provisional image, when developed, will form the final image in Platinum.

Reduction
Any chemical reaction which involves the gain of one or more electrons by an atom or an ion.

Sensitizer
A chemical solution which when applied to a support, such as paper, and allowed to dry, becomes light sensitive.

Sensitometery
Literally defined as sensitivity measurment. More specifically defined as a method of measuring the "speed" or light sensitivity of a photographic emulsion or sensitizer.

Silver Halide
A silver salt formed by mixing silver metal with one of the halides, i.e., iodide, chloride, or bromide. This silver halide compound is then suspended in a gelatin to form a light sensitive emulsion.

Sodium Citrate
Chemical Formula: Na3(CO2)CH2C(OH) (CO2)CH2(CO2) The sodium salt of Citric Acid. One of a number of chemicals that may be used as a developer for Platinum Prints.

Sodium Chloropalladite
Chemical Formula: Na2PdC14 The compound most commonly used as the source of Palladium for Palladium or Platinum/Palladium printing.

UV Radiation
The portion of the electromagnetic spectrum comprised of three regions: 1) near ultraviolet (4000-3000 A) which is present in sunlight; 2) middle ultraviolet (3000-2000 A) which is not present in sunlight; and 3) far ultraviolet (2000-100 A) which is not transmitted through air. The standard sunlamp emits UV radiation primarily in the near ultraviolet region.

Notes

55 Notes